BIOLOGY'S BUILDING BLOCKS

(Original French title:
Les biotechnologies)

by
Marie-Françoise Chevalier-Le Guyader

Translated from the French by Albert V. Carozzi and
Marguerite Carozzi

New York • London • Toronto • Sydney

First English language edition published in 1989 by
Barron's Educational Series, Inc.

© 1987 Hachette/Fondation Diderot-La Nouvelle Encyclopédie, Paris, France.

The title of the French edition is *Les biotechnologies*.

All rights reserved.
No part of this book may be reproduced in any form,
by photostat, microfilm, xerography, or any other
means, or incorporated into any information
retrieval system, electronic or mechanical,
without the written permission of the copyright
owner.

All inquiries should be addressed to:
Barron's Educational Series, Inc.
250 Wireless Boulevard
Hauppauge, NY 11788

International Standard Book No. 0-8120-4212-3

Library of Congress Catalog Card No. 89-6948

Chevalier-Le Guyader, Marie-Françoise.
  [Biotechnologies. English]
  Biology's building blocks / by Marie-Françoise Chevalier-Le
Guyader; translated from the French by Albert V. Carozzi and
Marguerite Carozzi.
    p. cm. — (Focus on science)
  Translation of: Les biotechnologies.
  Includes index.
  Summary: Describes the history and applications of biotechnology, discussing genetic engineering, the properties of cells and enzymes, and future applications in agriculture, industry, health, and medicine.
  ISBN 0-8120-4212-3
  1. Biotechnology—Juvenile literature. 2. Cells—Juvenile
literature. 3. Enzymes—Juvenile literature. [1. Biotechnology.]
  I. Title. II. Series.
TP248.218.C4413 1989
660'.6—dc20
                                                    89-6948
                                                    CIP
                                                    AC

PRINTED IN FRANCE

901 9687 987654321

Contents

Fermentation: Ancestral Biotechnology — 4
   Pasteur and the Beginning of Microbiology — 10
   Food Processes — 14
   Health Care — 16

The Cell: A Very Complex Factory — 20
   Unity and Diversity of Cells — 20
   The Cell: A Self-made Factory — 25
   The Cell and Biotechnology — 26

Enzymes: Molecular Workers — 30
   The Properties of Enzymes — 32
   Enzyme Engineering — 34

Applied Biotechnology — 40
   Genetic Engineering — 46

The Present Forecast of the Future — 50
   Toward the Control of Animal Reproduction — 50
   Custom-made Plants — 53
   A New Gastronomy? — 57
   Against Malnutrition — 58
   New Molecules for Medicine — 60
   New Sources of Energy — 62
   How to Use Waste Products — 63
   New Antipollutants — 66

Promises and Limits — 68
   An Economic Panorama Turned Upside Down — 68
   To Change the Genome — 70
   Biotechnology in Space — 72

Glossary — 75

Index — 77

Fermentation: Ancestral Biotechnology

Together with engineers, scientists are today working on new industrial techniques, called biotechnologies, in which the vital activities of living organisms are used as tools.

Such activities as domesticating animals and plants by breeding and agriculture date back several millennia. These foundations for the development of our civilization are, so to speak, ancestral biotechnologies.

However, major discoveries in biology are today revolutionizing the traditional ways of agriculture, of breeding, and more recently of industry. It all started in 1822 when Louis Pasteur was born in Dole, a small village in the Jura Mountains.

Water, salt, bread, and wine represent staple traditional foods in Mediterranean civilizations. Water and salt are inorganic mineral components; bread is said to be of organic origin because it results from the transformation by fermentation of a natural living product, wheat (flour). Such is also the case for wine, a fermented beverage derived from grapevines (grapes).

Humans have known how to make bread and wine for a long, long time. The Bible says that after the deluge, "Noah, the farmer, started to plant grapevines" (Genesis, IX:20). Bread is supposed to have been invented by the Hebrews. In the fashion of the Tuareg tribes, Hebrews baked cakes of dough made from flour, water, and salt

By 4000 B.C., the Assyrians already drank beer! The people of ancient Gaul called it cervoise. In today's breweries, fermentation of germinated barley is made in large mash tuns.

on hot stones in the desert. A saying goes that a cake was forgotten before being baked and that it was noticed later that it had swollen and that, after its baking, it was tastier, lighter, and more easily digested: bread was born.

Thus humans have used fermentation, which transforms products of organic origin in a natural way: this process is today called first-generation biotechnology.

When Pasteur was born, nobody knew why grape juice turns into wine and wheat into bread. Nothing was known about the making of cheese or yogurt from milk. These fermentations are inevitable processes: everywhere in the world where wine grows, traditions for the making of grape juice have never developed, and for a very good reason! Exposed to air, grape juice ferments and turns immediately into wine. Its sug-

Today a biochemist is able to explain the procedures of a baker. After kneading, yeast degrades starch by fermentation. The heat of the oven caramelizes the crust and expands carbon dioxide, thus creating holes.

THE BIOCHEMIST

protein	water	starch		yeast
		dextrin + glucose		
				Fermentation
protein	water	starch	dextrin + glucose	carbon dioxide + alcohol
			caramelization	expansion of gas / evaporation disappearance
soft part of the bread		caramelized crust	holes in the soft part of the bread	

ry flavor changes into an alcoholic flavor. This fermentation (called alcoholic fermentation) occurs when sugar is transformed into a particular alcohol, ethyl alcohol. For this reason, our wine-growing ancestors discovered the ways and means to control these inevitable fermentations and to produce wines with various flavors and for long preservation.

In fact, fermentation transforms food, favors its preservation, and gives it greater nutritional value.

Let us consider soft white cheese, for example. Milk exposed to air is not preserved. It "runs" or curdles and changes into curds and sour milk. This change in texture and taste is due to lactic fermentation. An acid called lactic acid appears in the milk and causes the milk to coagulate exactly as would happen if a few drops of lemon juice were added. The curds, eaten as they are or used to make cheese, are very similar to the regurgitation of a baby a few minutes after its bottle. The milk has

Wine and other products of grapes

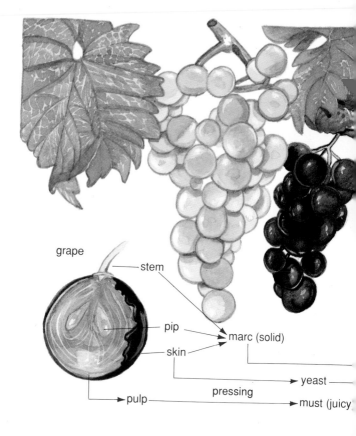

grape — stem — pip — skin — pulp → marc (solid) → yeast → must (juicy) — pressing

White wine: To obtain white wine from black grapes, the skins and branched stem portions of the fruiting cluster—which contain the coloring agents—must be removed from the must. "Blanc de blanc" is made with white grapes.
Champagne: Sugar and yeast are added to dry white wine after bottling. The fermentation of sugar produces an important release of carbon dioxide: champagne becomes bubbly.
Alcohol: Distillers remove the marc and let fermentation continue: yeasts produce alcohol, which is later separated from marc by distillation in a still.

```
                        ──────────────► red coloring + flavor
              fermentation                        │
                        ──────────────► alcohol ──┴──► red wine
              carbon dioxide
```

Vinegar: Some bacteria transform ethyl alcohol into acetic acid by means of fermentation. These bacteria develop in appropriate vats at the surface of wine and form a thick gluey skin which is actually the mother of vinegar.

Grape Juice: If fermentation of the must is prevented, nonalcoholic grape juice is produced. The process consists of washing and treating grapes so that developing yeasts are eliminated. Grape juice is kept cool as is pasteurized.

thus curdled in the baby's stomach where digestion started. Therefore, soft white cheese is, so to speak, predigested food and is easily assimilated by the organism.

Every country has its traditions. For instance, milk produces yogurt in Bulgaria and *kefir* in Turkey. In France, children eat preferably one of the many brands of soft white cheese, such as Brie, and in America, cream cheese is a favorite with children and adults.

Similarly, fruit and grains ferment and turn into alcohol: cognac (from grapes), sake (from rice), and whiskey and beer (from barley).

Pasteur discovered that the various fermentations are caused by microscopic organisms, called microorganisms or microbes, which are just as alive as the stem of a vine. Our ancestors were thus practicing biotechnology without knowing it since they took advantage of the properties of these microorganisms, which are true biological tools.

These microorganisms exist in a great number and variety and include bacteria and microscopic fungi, among which yeasts are the most abundant.

Pasteur and the Beginning of Microbiology

The empirical knowledge of fermentation became scientific knowledge with Pasteur.

He knew about the analyses by Louis Gay-Lussac who had studied the chemistry in alcoholic fermentation several years earlier. Pasteur believed that the explanation of fermentation given by his contemporaries was not complete.

The only tools Pasteur owned were a drying oven and a microscope. However, because of his great scientific rigor, he found an explanation common to all these phenomena, which seemed to be unrelated to each other.

He observed fermented substances under the microscope and sketched the tiny globules taken from the gray areas floating at the surface of sour milk. He saw them bud and multiply. When he injected a drop of these globules into fresh milk, he caused immediate lactic fermentation.

Were living organisms therefore responsible for fermentation? Pasteur worked for more than 20 years to convince his

Louis Pasteur (1822-1895) is buried in a mausoleum located in the basement of the Institute Pasteur of Paris, which he had built and where he had lived. A museum may be visited there.

contemporaries. First, he had to prove that these microorganisms existed before he could demonstrate what they did.

He invented spherical glass containers with long swan necks (retorts) and filled them with a broth called culture medium favorable for the development of microorganisms, which had been eliminated beforehand through boiling. These retorts are in contact with the outside air through their bent neck. Gases pass through, but microorganisms are deposited in the curvature of the neck without reaching the liquid, which remains unchanged. To start alteration of the liquid, the retort must be tilted until the liquid mixes with the dusts in the neck. It is said that the medium has been "cultured" with microorganisms.

Pasteur worked in the basement of the observatory of Paris to find an atmosphere rich in microorganisms. He went to

Cheese

Every region or country makes its cheese from local milk: from a cow, a sheep, a goat, a reindeer, a buffalo, an ass, or a camel.

The techniques resemble each other. First, milk is coagulated (curdled) with rennin (an enzyme from the stomachs of young ruminant animals) or with an acid produced by the fermentation of lactic microorganisms at the surface of the milk.

Thereafter, the curds are drained. In France, for instance, soft white cheese is made from curds.

Curds may also be transformed by microorganisms during fermentation. A shepherd, having forgotten his soft white cheese in a cave where he had taken shelter, found it later a little, moldy but delicious : thus Roquefort cheese was born.

Certain microorganisms transform the paste of curds into blue cheese or Roquefort. Others act upon the crust such as Camembert and Brie. Washing the crust with saltwater makes Pont-l'Evèque and Münster . It is said,"there are as many kinds of cheeses as there are wines." It should be added, "as there are microorganisms."

the Alps to collect clean air (without microorganisms), climbing to the top of a glacier, the "Mer de Glace" at Chamonix, with twenty retorts! He observed that they were not "cultured."

Little by little, Pasteur proved first that milk, grape juice, germinated barley (malt), hydrated flour, and the juice of the sugar beet ferment only after having been in contact with microorganisms carried by the air.

Second, he demonstrated the role of these organisms. Without oxygen, they decompose or—to feed themselves—they partially degrade organic molecules, in particular glucose. Fermentation is therefore a partial degradation, and the end product (such as ethyl alcohol and lactic acid) changes the texture and the taste of food. These microorganisms are also able to live in the presence of oxygen. In this case, they completely degrade the initial molecules: this is respiration.

Clearly, fermentation provides less energy for microorganisms than respiration. Fermentation resembles a fire with incomplete combustion. However, fermentation is able to culture in a great number o environments that lack oxygen

This is how Pasteur initiate microbiology, a science tha studies the life of microorgan isms.

After his discovery of thei role in fermentation, Pasteu had only one wish: to extend h research into the domain c health. Together with his stu dents (Calmette, Chambe land, Duclaux, Laverar Metchnikov, Nicolle, Rou and Yersin), he showed tha bacterial microorganisms ar responsible for many anima and human diseases fatal at tha time: the disease of silkworm anthrax, yellow fever, choler plague, and diphtheria.

These discoveries had grea implications for medicine.

Food Processes

1. *Sterilization*, or how to elim nate all microorganisms in culture medium, is perhaps tl most basic technique used microbiology. As the farm who plows a field and remov all weeds before sowing whe or corn, a microbiologist r moves from the culture mediu all microorganisms by heating at high temperature (more th

100° C [212°F]). This is sterilization. Once sterilized, the medium is cultured with a colony of bacteria using instruments that are also sterilized. A colony of bacteria is a mass of perfectly identical bacteria.

Soon after its invention, sterilization was used in agribusiness industries for the preservation of food. Canning partly replaced the traditional methods of preservation in brine or vinegar. Today, the method has been improved for milk and cream. UHT sterilization (ultrahigh temperature) is used. It consists of heating products at 200 or 230°C (360°F or 414°F) temperature for a very short time. Thus the taste is less altered.

2. *Pasteurization* is how to destroy only part of the present microorganisms in a medium undergoing fermentation.

In the 1860s, French beer manufacturers (brewers) were alarmed because their beer turned and became unfit for drinking. How was it possible for the French brewers to compete with German brewers? Pasteur was called in to help. To

Fermentation and respiration

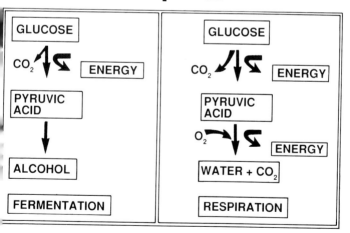

During fermentation *the degradation of glucose is incomplete and produces little energy.*
During respiration *the degradation of glucose is complete. It produces much energy but requires oxygen.*

find the cause of the spoilage of beer, Pasteur visited many breweries and studied their manufacturing methods. He even went to England to observe worts of beer!

How is beer manufactured? Germinated barley, called malt, is rich in sugars, colorings, aromas, and proteins. After fermentation, it changes into wort (as wine changes into must). It is then boiled with hops, which give beer its bitter taste. The brewer then produces two successive fermentations during which sugars disappear while bubbles and froth appear.

Unfortunately, parasitic microorganisms introduced either by machines, air, or raw materials culture and then alter the wort, that is, beer in the process of being produced. The mash tuns must be emptied, and fermentation must start all over again with another yeast.

Pasteur proposed to heat the beer at a temperature of 142-145°F to eliminate most of the harmful microorganisms. This method is is known as pasteurization. Pasteurized beer was born. Thereafter, wine, milk, cider, and grape juice were also pasteurized for commercialization and export.

Health Care

1. *Asepsis and antisepsis* are radical cleaning processes.

The importance of a discovery is sometimes matched only by its simplicity and its immediate efficiency. The discovery of microbes, in particular those causing disease (pathogens) led to an immediate conclusion: when a bacterial infection is feared, one tries first to destroy bacteria by antiseptics such as hydrogen peroxide, alcohol, or Dakin's solution (an alkaline solution of sodium hypochlorite).

Furthermore, in surgery where the most aseptic possible environment is required, surgical instruments are sterilized and the surgeon and the nurse wash their hands and wear perfectly clean gowns. This aspect seems quite common today however, a century ago, a surgeon wore his everyday clothes during surgery, and the scalpel might have been used beforehand for dissecting corpses! At that time, the risk of infection was the greatest danger Pasteur's discovery has th

Surgeons *sometimes operate under tents that airtight and therefore perfectly sterile. T handle instruments wearing long gloves.*

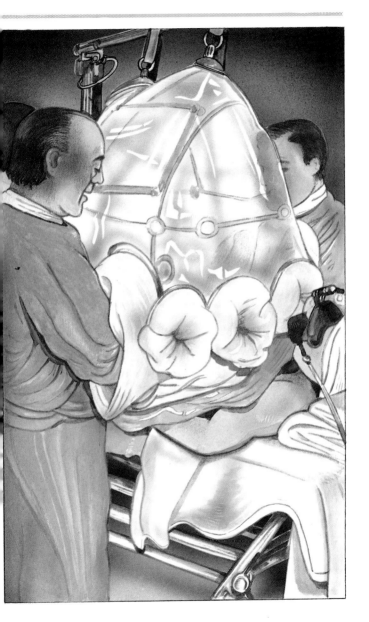

Penicillium *is a miroscopic fungus, the producer of penicillin, the first antibiotic discovered. Industrial production occurs by the extraction of the fungus from media in large bioreactors.*

saved many human lives by asking surgeons to wash their hands.

2. *Vaccination* is how to fight against the bacteria that invade an organism.

A vaccine against a disease consists of the pathogen whose virulence has been weakened by appropriate physical or chemical treaments. Thus an organism does not catch the disease because the injected microorganism is no longer active but reacts by developing defense barriers against this new host. If the microbe attacks, the counterattack occurs immediately because the defense weapons are ready.

For this reason, we are vaccinated against several diseases:

tuberculosis, diphtheria, tetanus, whooping cough, and polio. Such systematic prevention has improved human health and extended our life span.

In 1885, Pasteur achieved the first really scientific human vaccination on a young man from Alsace bitten by a dog that was infected with rabies. The vaccine consisted of extracts of the infected dog's nervous system, extracts that were attenuated (weakened) and treated chemically to render inactive the microbe they contained. The vaccination was successful, and since then thousands of persons throughout the world have been kept from developing rabies by being given the Pasteur vaccine after having been exposed to rabies. The works by Pasteur and his students Roux, Chamberland, Metchnikow, Duclaux, and Yersin created such enthusiasm that a national subscription and outside donations financed the building of the present Pasteur Institute. After the vaccine against rabies, others were made: against diphtheria in 1890 and tetanus in 1893. These were the beginnings of immunology, a science that studies the defense mechanisms of organisms, a science to which is still attached the name of the Pasteur Institute.

3. *Antibiotics* cure diseases caused by bacteria.

Pasteur and a student, J. F. Joubert, showed that certain bacteria inhibit the growth of one of their fellow creatures causing anthrax. Some microorganisms are thus able to inhibit growth in others by secreting particular substances called antibiotics. The first among these substances was isolated in 1896, and others followed. However, the chemical importance of these molecules was understood only much later in 1929 by Alexander Fleming. He showed that a fungus, *Penicillium*, secretes a compound selectively able to inactivate a wide range of bacteria without induly influencing the host, a substance he named penicillin. Its use as a drug against bacterial diseases revolutionized the practice of medicine during World War II.

Every microorganism synthesizes a different antibiotic. Each species lives in a specific environment and chases its competitors with killer molecules—antibiotics! This is one of the laws in the life of microbes.

The Cell: A Very Complex Factory

Beginning in the middle of the seventeenth century, after the discovery of the microscope, naturalists tried to understand the mysteries of the organization of plant and animal tissues. One of them, the English naturalist, Robert Hooke (1635-1703), observing cork, found that it consists of cavities adjacent one to another as are cells in a honeycomb. He called them "cells" from the Latin cellula, that is, "small rooms."

Microscopes improved and observations continued. In the 1830s, two German naturalists, Theodor Schwann (1810-1882) and Mathias Schleiden (1804-1881), suggested that the cell is the smallest autonomous unit capable of showing signs of life.

The smallest living things therefore consist of one cell (bacteria, ameba, and paramecia) and are called unicellular. More compex organisms consisting of numerous cells are called multicellular. Many billions of billions form a human being.

Unity and Diversity of Cells

At first sight, cells appear very different one from the other. What similarity is there indeed between a nerve cell and a muscle cell?

Nevertheless, the same general pattern of organization is found in each and every cell. They all have in common the following characteristics:

The largest part of a cell con-

In the tissues of the nervous system of our brain, cells called neurons appear all tangled up. However, this confusion is only apparent and hides a functional and complex organization.

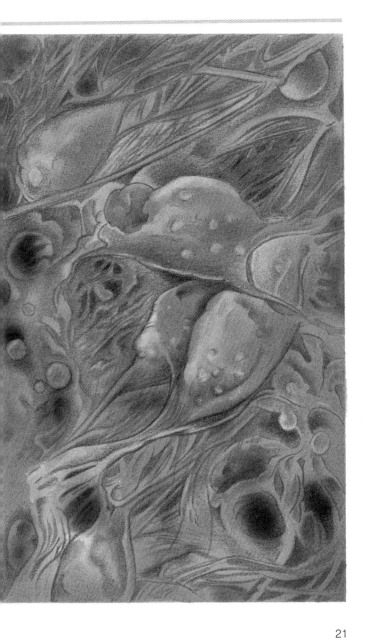

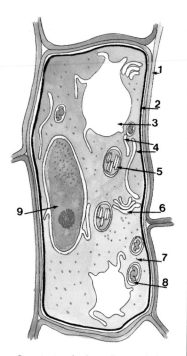

Organization of a plant cell: *1. Rigid cellulose wall 2. Cytoplasmic membrane 3. Vacuole 4. Endoplasmic reticulum 5. Chloroplast 6. Golgi appartus 7. Ribosome 8. Mitochondrion 9. Nucleus 1 and 5 are characteristic of plant cells.*

sists of the cytoplasm: this is where the majority of chemical reactions takes place.

The separation between the internal cellular environment and the external one is represented by the cytoplasmic membrane.

Small constituents visible under the microscope, called cytoplasmic organelles, are scattered in the cytoplasm. They play very specific roles:

In the mitochondria take place the phenomena of respiration. Mitochondria represent the power plants of the cell. Therefore, fermentation occurs in the cytoplasm.

Chloroplasts are organelles characteristic of plant cells. They produce photosynthesis by converting carbon dioxide and water into sugars in the presence of sunlight. Chlorophyll, the coloring matter contained in chloroplasts, gives them as well as plants a green color.

Other organelles have other specific functions. For instance, lysosomes destroy those parts of the cytoplasm that are no longer of any use or they digest large particles taken from the outside. The Golgi apparatus helps to secrete molecules produced by the cell in its cytolplasm.

The endoplasmic reticulum irrigates the entire cell and serves as communication system in the cytoplasm. It also represents an exchange surface with the outside.

A nucleus coordinates all cellular activities. It is the

"control tower" of the cell. It contains particular molecules: nucleic acids, such as DNA (deoxyribonucleic acid), instrumental for genetic information. How is that information transmitted to the entire cell?

Inside the cytoplasm, other small organelles are found, which are called ribosomes. They synthesize proteins, the building blocks of living matter. During synthesis, the required information comes from the nucleus, or more precisely from the DNA. Indeed, the blueprint of proteins is written into this very important molecule. However, DNA never leaves the nucleus! To overcome this obstacle, a copy of DNA is made by another nucleic acid, the RNA (ribonucleic acid).

It is thus this copy, RNA, which carries the information of DNA from the nucleus to the ribosomes. When its message is delivered the copy is destroyed.

Hence, the following analogy is often presented. The nucleus resembles a library where valuable books cannot be checked out. The RNA is the analog of a photocopy that does not destroy the original but can be discarded without any problem after its use.

Thus, whoever wants to make protein must have the blueprint of the DNA. This is the principle of one of the techniques most frequently used today for the study of cells, genetic engineering, as we see later.

Finally, let us not forget one major difference between plant and animal cells other than the existence of chloroplasts. The plant cell is surrounded by a rigid wall consisting mainly of cellulose. This wall prevents important deformations of the cell and consequently of all movement of the plant. This cellular characteristic causes the quasi-immobility in the plant kingdom.

We have briefly described the major cellular components of plants and higher animals. However, a number of simple cells exist, represented mainly by bacteria. These cells have no well-defined nucleus surrounded by a nuclear membrane. A unique molecule of DNA swims directly in the cytoplasm. For biologists, this means that it is much easier to reach this molecule. These are the cells that are indeed used commonly for the development of biotechnologies.

Photosynthesis

From daily experience, humans have learned that every construction (including manufacturing and synthesis) needs raw materials and energy. In photosynthesis, these materials are carbon dioxide and water: the end product is sugar and, through its help, all the components of a cell. Energy is provided by sunlight. Organisms able to realize this synthesis are called photosynthetic.

Only plants and certain bacteria can photosynthesize in their cytoplasm; they have organelles, called chloroplasts, in which photosynthesis occurs. These chloroplasts are in fact very complex solar mini-power plants.

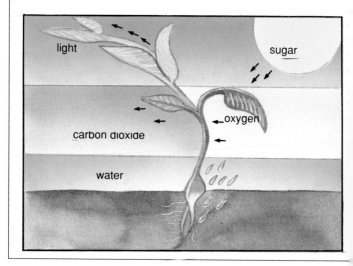

Nevertheless, the conclusion should not be reached that all cells are identical—on the contrary! By varying the characters in a quantitative or qualitative fashion, it is possible to create from an identical blueprint an extraordinary diversity of cells. Their size ranges from a few micrometers (7 μ in diameter for the red human hemoglobin molecule) to several centimeters (7.5 cm, or 3 inches, for the ovum of an ostrich, that is, its egg yolk). Shapes also vary: some cells have the shape of a sphere, a spindle, a spiral, a rodlet, or a tree.

Some cells can change their shape to move, to find food, as ameba do. Chemical compositions may present unusual characteristics: fragrance of flowers, the coloring of fruits and petals, the existence of proteins capable of contracting in muscle cells, and the storage of fat.

Thus, starting from a single "egg cell," an organism as complex as the human body may be harmoniously constructed with the help of numerous specialized cells. It is during the development of the embryo that cells differentiate according to yet not too well known processes.

The Cell: A Self-made Factory

If a medium is cultured with one bacterium only, very soon millions of them exist. Indeed, under favorable conditions of temperature and food, each cell divides about every 20 minutes. It can be observed that every daughter bacterium is completely identical to the mother bacterium, the common ancestor. The million bacteria thus formed resemble each other like twins! In biology, these are called *clones*.

To grow, cells absorb the needed components found in the medium. To achieve this with a minimum effort, cells are able to adapt their enzymes to the nature of the molecules of their external medium. Enzymes are, as we see in the following chapter, the workers of the cell.

A particularly famous example is that of lactose. This is the sugar of milk, with two small molecules attached to each other: glucose and galactose. Since the molecule that provides the bacteria with energy is glucose, lactose must be broken up—or hydrolyzed—to separate glucose from galactose.

What happens when bacteria find both glucose and lactose in the culture medium? First, glucose is absorbed and lactose remains untouched. The cell follows the law of minimum effort: it first takes from the immediate environment what it needs.

However, when free glucose is completely used up, the only source available to the bacteria is lactose. What do they do? They produce a specific enzyme, betagalactosidase, which hydrolyzes lactose in the medium so that glucose is freed. Furthermore, another specific enzyme (an isomerase) can transform galactose into glucose. Paying the price of some effort, namely, the synthesis of adaptable enzymes, bacteria have found a new source of glucose.

What is remarkable in this phenomenon is that betagalactosidase can be observed in the cell only when lactose is present in the medium. Hence, bacteria adapt their enzymes to their living conditions. This is of course possible only because the blueprint of the enzyme exists in the DNA of these bacteria. However, another source of glucose, such as cellulose, is not available because the DNA of these bacteria do not have it blueprint. If it were artificially introduced into the cell, these bacteria would be able to assimilate cellulose.

The Cell and Biotechnology

Those who practice biotechnology must learn about cellular functions. Indeed, the following four factors should be considered:

Isolated cells are the easiest and most suitable for cultures: bacteria, yeasts, algae, plant cells, and so on.

Enzymes are the workers that we want to use. We have seen that their activity depends upon the environment and the needs of the cell.

To change the enzymatic outfit of a cell requires action at the level of the DNA. This delicate operation is of course only possible with a relatively small number of cells.

If it were not possible to obtain clones, genetic engineering would have no effect. A clone that multiplies new characteristics as much as needed can actually be considered for industrial purposes only when isolated cells are used.

Every living being produced by sexual reproduction consists of a single cell only at the beginning of its life. This is the *egg cell*. Only at the beginning of its life—not later—can it be manipulated.

Deoxyribonucleic acid or DNA. *The molecule of DNA consists of two strands coiled in a spiral around the same axis, thus forming a double helix. A copy of the genetic code is made in the nucleus by enzymes that locally spread apart the strands and synthesize a molecule of RNA.*

Life molecules

Living organic matter consists of water, proteins, carbohydrates, lipids, vitamins, and salts. **Proteins** *are gigantic molecules, or macromolecules, formed by a chain—of variable size—of small molecules: amino acids. Twenty-one different amino acids exist. The number of possible combinations to make one protein is thus virtually unlimited. This explains the diversity of proteins and the many roles they play in the life of a cell: structural proteins for the construction of cells; defense proteins (anitbodies) against invaders foreign to the body; chemical proteins (enzymes), which are the workers of the cell; and messenger proteins (hormones and neurotransmittors), which control the activity of cells and organs.*

Carbohydrates*, more commonly called sugars, form, together with lipids, energy supplies for animals and plants. A first category of carbohydrates, soluble in water and rapidly assimilated by the body, are simple sugars, such as glucose (sugar of animal cells), fructose (sugar of fruit, vegetables, and honey), and other more complex sugars, such as saccharose (sugarcane or sugar beet), lactose (milk sugar), and maltose (sugar after transformation of starch and dextrins). A second group of carbohydrates, insoluble in water and assimilated slowly by the organism, are such macromolecules as starch (reserve sugar of grains), dextrins (degradation product of starch), and cellulose (main component of the walls of plant cells). All carbohydrates are formed by numerous chains of simple sugar molecules.*

Lipids, also called fats, are molecules that are insoluble in water. They are the principal components of cell membranes and form an important energy supply for the organism. Several families with a complex chemistry are distinguished. Among them are glycerides, composed of a type of alcohol; glycerol, and of very long molecules: fatty acids.

Finally, **vitamins** are indispensable molecules for the organism in very small amounts. Several families exist, named A, B, C, D, and so on. Lack of vitamins causes a condition named avitaminosis. For instance, a deficiency in vitamin D causes rachitis and in vitamin C, scurvy.

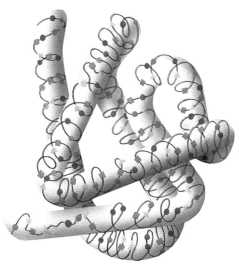

A protein *is an assemblage of chains of amino acids.*

Enzymes: Molecular Workers

Enzymes are perhaps the most unusual molecules of living matter. A comparison of the functions of a cell with those in a factory shows the role of enzymes.

The smooth running of a factory requires raw materials (for instance iron for rails), energy (in the form of electricity to run machines), tools, and workers. The latter actually run the factory.

In a cell, raw materials are complex: mineral salts, proteins, sugars, and/or fats depending upon each case. Energy is obtained during partial degradation (fermentation) or total degradation (respiration) of sugars and fats. The pair worker-tool is represented by particular proteins, namely enzymes.

To understand the functions of enzymes, let us look at the example of the use of a food supply by a germinating grain of wheat. While attached to the ear, the grain gorges itself with sugar. This sugar is at the beginning glucose, a small water-soluble molecule identical to the sugar in our blood. However, considering that a cell consists of 90% water, it is preferable to render it insoluble for easy storage. This requires a particular procedure: molecules of glucose are attached to each other like pearls in a necklace and form a "supermolecule." This new molecule is called starch. It is so big that it becomes insoluble. Starch is found in many foods of vegetable origin, including ce-

In a biochemistry laboratory, minicomputer-controlled instruments have replaced Pasteur's retorts.

reals, potatoes, bananas, and beans. In chemistry, these large molecules are called polymers.

What happens when a grain of wheat germinates? The young plant needs sugar because it spends a lot of energy to grow. However, it needs it in the form of glucose, not starch! This is the time for an enzyme to intervene: it divides the starch molecule and uncouples as many molecules of glucose as needed. In this particular case, the enzyme is called amylase. It resembles very much a worker who starts working with a tool when called upon.

An amylase of the same type occurs in our gastric juices and allows the digestion of starch. Indeed, digestion represents a molecular simplification that breaks down large molecules ("macromolecules") into smaller ones that are directly assimilated by the organism because they can pass into the bloodstream.

What happens if we do not have the necessary enzyme? Macromolecules are not digested and pass through the digestive tract and remain intact. Hence, cellulose—a very common molecule in the plant kingdom—is not digested by our digestive tract because we do not have the enzyme cellulase that would break down cellulose in our gastric juices.

The Properties of Enzymes

All reactions of the cell are carried out (in scientific terms, catalyzed) by enzymes. However, there are thousands of reactions in a cell! Can the same enyzme act upon several reactions? Such a question brings us to the study of the characteristics of enzymes.

In fact, an enzyme is a most specific worker that always carries out the same reaction on the same molecule. To understand this, let us first look at the principal functions of enzymes.

As we have already learned, some enzymes are excellent "scissors." Every molecule finds, if necessary, an enzyme called hydrolase that degrades or breaks down this molecule. This property has been used for the manufacturing of biological detergents with the help of the so-called gluttonous enzymes (amylases), which are able to destroy molecules of fat and blood, for example, and to remove stains.

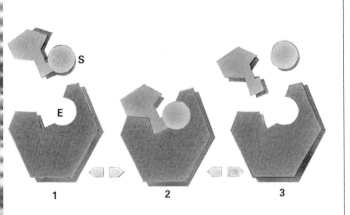

Why Are Enzymes So Efficient?

Enzymes are proteins and therefore have a precise spatial arrangement.

The enzyme's concave side (area E) has the same shape as the area of the substrate to be catalyzed (area S) [1]. As a key in a keyhole, the substrate fits into the enzyme.

Thus a compound enzyme-substrate is formed, filled with the energy of the enzyme [2].

This energy is necessary for the reaction of catalysis.

Thereafter, the compound enzyme-substrate falls apart: the substrate is unlocked from the active site, and the freed enzyme is ready for a second substrate [3].

Thus, it does very rapid chainwork and is able to catalyze easily several thousands of molecules.

Other enzymes are excellent "staplers": they can build a new, larger molecule by linking together two molecules. These enzymes are called synthetases.

Some enzymes can rearrange a molecule by changing the order of links between certain parts of that molecule.

Others can transform a molecule into its mirror image. In chemistry symmetrical molecules are called isomers. Therefore, these enzymes are named isomerases.

It is easy to understand that, given all these possibilities, the molecular transformations in the cell are extremely numerous.

After these few examples, let us repeat the properties of enzymes. Biologists say they have a double specificity of reaction and of substrate: of reaction because they can only catalyze one type of reaction (for instance, a "stapler" can never do the job of "scissors"); of substrate because they act only on the same type of molecule (amylase never breaks down cellulose).

Therefore, when studying an enzyme, one must know two important things: what kind of reaction it catalyzes and on which substrate.

Enzyme Engineering

The German chemist Otto Roehm first introduced the use of enzymes in industry. He noticed that tanners used a rather strange method to treat leather to make it pliable: they macerated hides in dog dung! Roehm succeeded in isolating the protease present in the digestive tracts of dogs. Protease degrades the proteins in hides and renders them pliable. Roehm sold these enzymes because they were much less offensive to work with.

Roehm thus created the first enzyme factory in the world. Later, he produced a great variety of enzymes for leather, textile, and pharmaceutical industries as well as for the making of fruit juices.

Today, more than 2000 different enzymes are known that are capable of producing chemical changes in substances. However, only a few have an industrial use. These are essentially:

Proteases degrade proteins. They are used in detergents.

Amylases degrade starch molecules. They are used in bakeries and breweries.

Isomerases transform glucose

An Enzyme Bioreactor

An enzyme bioreactor is a container in which biochemical reactions are catalyzed by enzymes. There are various immobilization techniques of enzymes (1) in a bioreactor:

by immobilization onto a porous support, enzymes are trapped in the cavities of a synthetic sponge (3).

by microencapsulation, enzymes are enclosed in porous microscopic spherical membranes (4).

by covalent coupling, enzymes are attached to solid supports of glass or plastic material by chemical bonds (2).

by entrapment, enzymes are fused together by means of chemical substances that are a kind of gel (5).

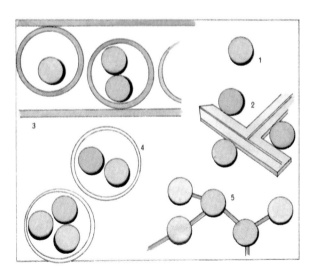

A New Source of Sugar: Corn

Traditionally, sugar is extracted from sugarcane. Until the Age of Napoleon, France imported sugarcane from the West Indies, which were an important producer. However, because of the British blockade at the beginning of the nineteenth century, the cultivation of the sugar beet was begun in northern France, and sugar has been produced there in part ever since (for the French market).

The history of sugar is not finished, however, because beginning in 1975, a third natural source of sugar became popular: corn. Sugar, in the form of granules or in cubes, consists of saccharose (sugar of plants). It is extracted from sugarcane or

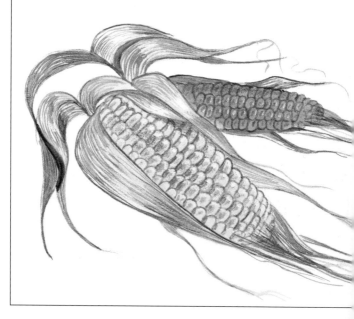

sugar beets and then refined.

A grain of corn, however, contains starch. It can be transformed into a sweetening product formed by another sugar, fructose. This requires two successive bioconversions: from starch into glucose and then from glucose into fructose. Enzyme immobilization techniques permitted industrialization of these conversions. This was the first great victory of enzyme engineering. A sugary syrup is produced that is used in agribusiness industries. It sweetens, in particular, Coca-Cola, fruit juices, and other beverages.

The development of this technique allowed the United States and Japan to produce sugar without depending upon sugar-producing countries during the sugar crisis of 1974.

into fructose. They produce molecules with a taste of sugar from corn, not only from sugarcane or sugar beet.

Rennin curdles milk.

Lactases break down lactose, the sugar of milk.

These enzymes have revolutionized traditional industries based on fermentation: breweries, bakeries, and wine making.

Now the brewer uses amylases to act directly on raw barley grains instead of going through the process of adding malt (germinated barley).

The baker also tends more and more to add amylases to the dough. They accelerate the breakdown of starch and the action of yeast: the dough therefore rises much faster.

A wine maker needs enzymes to clarify the wine; the enzyme pectinase breaks down the jelly (pectin) in grapes.

The general use of enzymes is hampered by one of their biological properties: their fragility. This is why research is being done today to increase their life span. Enzyme immobilization techniques appear very promising in that respect, by immobilization onto solid supports or by entrapment inside a matrix or gel that is permeable to the enzyme. Such immobilization considerably lengthens the life span of enzymes.

They can then be reused several times to act upon substrates that we want to transform. Furthermore, since the enzyme is protected, it can be used at higher temperature, which allows an increased speed of reaction. Hence containments are built where enzymes are immobilized and where their conditions of optimum action are assured. These are the so-called bioreactors.

All these steps require an active collaboration between chemists who prepare the supports and biologists who purify and study enzymes.

This is the reason that biotechnologic industries need large amounts of research funds.

Microencapsulation

In solution (figure on the left), enzymes have a very short life span. To be able to use them, this span must be extended.

One of the techniques used for this is microencapsulation. As seen on the figure on the right, enzymes are enclosed in tiny porous spherical membranes.

Thus, their life span can be doubled.

This is particularly necessary for industrial use.

Applied Biotechnology

Biotechnology would not exist without very sophisticated techniques. Some of these, which have become classics, pertain to the study of the cell (its chemical composition, its organization, its function, and its development). These are essentially techniques of separation and purification. However, biotechnology is largely based upon two recent discoveries: cell culture and genetic engineering. The first discovery obtains in vitro (that is, in the laboratory under special conditions) the replication of cells that may result, for instance, in the formation of a plant. The second discovery offers the possibility of modifying hereditary characteristics of a cell and monitoring the synthesis of certain molecules.

Starting at the beginning of the twentieth century, *in vitro* cultures of segments of animal and plant tissues were attempted. Research on animal tissues was soon crowned with success, whereas that on plant tissues failed for a long time.

In 1903, the French biologist Justin Jolly succeeded in the culture of red blood cells of a triton (a kind of mollusk). However, the pace of cell division soon slowed and finally stopped.

In 1910, another French biologist, Alexis Carrel, triumphed in what seemed impossible: he kept alive and growing indefinitely a culture of heart cells of a chicken embryo. His technique proved successful for the majority of cases, and numerous animal cells were rapidly grown in culture. One of the first important industrial applications of this technique was the preparation of a vaccine against polio.

1. Cutting a rosebud with sterilized instruments. 2. Development of the leaves of the bud in an artificial medium. 3. Change of medium to grow roots. 4. The shoot is taken from the medium and planted in a pot. 5. Flowering of the rosebush.

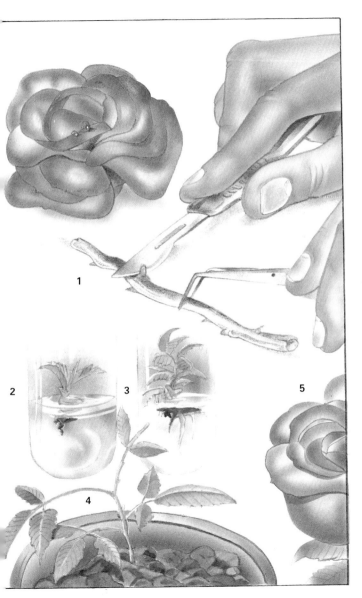

A piece of a palm leaf is grown in an in vitro culture. It produces leaves and "embryoids," which in turn are placed in culture and can produce entire plants.

Not before 1939 did it become possible to isolate small numbers of cells from certain plants and to keep them alive indefinitely with artificial cultivation. Successful plant cultures followed, at first slowly and then faster and faster: cultures of clumps of embryonic cells (meristem) in 1952, formation of small organisms resembling embryos, called embryoids, from isolated cells of carrots in 1958, plants regenerated from stamens in 1967.

Today, numerous experiments can be done:

Culture of isolated cells in suspension in a nourishing liquid is used when the cell prouces an important molecule of interest for commercial use. Thus cultures of microscopic algae consisting of a single cell that produces hydrocarbons and gels are easily obtainable.

Cells of more advanced plants are more difficult to grow, sometimes even impossible, because they lose their initial specificity. For animal cells, most techniques are now perfected.

Regeneration of plants from meristems is particularly useful in producing virus-free plants. Indeed, when removing the meristem, deeper tissues of the plant can be chosen where viruses are absent. Therefore, new plants are free of viruses. After a viral plant epidemic that otherwise would be a catastrophe, it is possible to rebuild crops relatively fast with healthy plants. This was done in France for the potato after the drought of 1976.

Embryoids can be formed *in vitro*. These can be produced fast and in large quantities.

Embryoids have the same capacities as seeds. Furthermore, in most cases it is possible to freeze them at the temperature of liquid nitrogen (-196°C [-353°F]) without damage. Embryoid banks preserve numerous varieties at low cost and in reduced space.

This technique has been used successfully for the culture of the coconut palm in Africa and Indonesia. From one tree alone, it is possible to obtain as many embryoids as desired. All trees regenerated from these embryoids have identical genetic characteristics and thus represent clones.

This is of considerable economic interest because one tree alone chosen for certain advantages can be multiplied by a thousand times. Unfortunately, these techniques cannot be applied to all plants; for those that are economically the most important, such as cereals (wheat, corn, and barley) and leguminous plants (beans, peas, and soybeans), the difficulties are very great. Much research is being done today to solve these problems.

Solar Biotechnology

During fermentation, yeasts, fungi, and bacteria are nourished by such substances as glucose, agricultural waste products, and paraffins. These raw materials are expensive and represent at least 50% of the production costs. With photosynthetic microorganisms, the problem disappears. Microscopic algae and cells of advanced plants need nothing but water, carbon dioxide, mineral sub-

stances, and sunlight to live! Some of these microorganisms secrete substances that are of interest to the pharmaceutical, cosmetic, and agribusiness industries, as well as energy, but their chemical synthesis is difficult and costly on a commerical scale. Therefore, techniques have been perfected to cultivate those organisms that live on a frugal diet. They are trapped in photoreactors—a kind of solar bioreactor—designed to catch as much solar energy as possible.

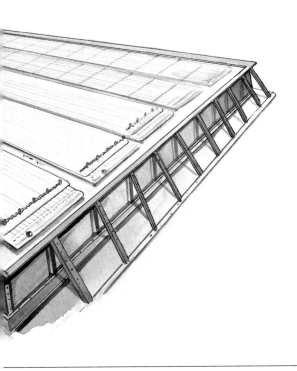

Genetic Engineering

This technique has spread the knowledge of biotechnology among the general public. It is indeed a spectacular technique because it offers potentially unlimited opportunities for bringing about new combinations of genes in animals and plants. This technique nevertheless appears a bit dangerous because no scientific obstacle seems to forbid its application to humans. Hence the question: science fiction or reality?

However, things are much less fantastic than they are often said to be. For the time being, only a very few genes (a gene is a transmitter of inherited characteristics) can be introduced at the same time into the same organism. For the present, a unique gene usually suffices for the task at hand. Manipulations of "science fiction" imagined by some would require a large number of genes, and this will be impossible for a long time to come.

Theoretically genetic engineering is simple, but experimentally it is quite complex.

We saw earlier that the blueprint of a protein is written into the DNA of the nucleus. If one wishes the synthesis of a particular protein by a microorganism or a more advanced organism, one need "only" insert that fragment of DNA coded for this protein into the DNA of the host cell.

However, here difficulties begin that are of various orders of magnitude:

Track down the fragment of DNA that one wishes to transfer into another cell.

Isolate and produce a copy or chemical synthesis of this fragment of DNA.

Find a vector or carrier system that, like a Trojan horse, is capable of entering a bacterium.

Insert this transformed vector into a bacterium.

Multiply the gene by multiplication of the bacterium; an operation called cloning.

Isolate genes now in great number.

Reinsert the gene by a new vector into the desired host cell.

Monitor the expression of the gene so that the transformed cell produces the desired protein.

Hence, there are three kinds of manipulations. First a fragment of DNA must be obtained. Second, it must be du-

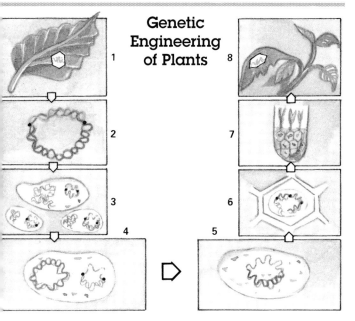

1. Extraction of the gene. 2. Insertion into a first vector. 3. Cloning. 4 and 5. Insertion into a new vector (a bacterium). 6. Insertion of the gene into the plant cell. 7. Formation of a shoot in vitro from the transformed plant cell. 8. Regeneration of a complete plant containing the gene in each cell.

plicated so that such fragments exist in large quantity. Third, the gene must be inserted into the desired host.

It is impossible here to give all the details step by step. However, scientists have followed these steps and were able in 1979 to produce human insulin, a polypeptide hormone, by means of a bacterium. Insulin was produced commercially in 1981, and this hormone, used for the treatment of diabetes, is thus made relatively cheap. Other protein hormones have since been produced by genetic engineering, for example a growth hormone.

We now understand that genetic engineering would have little impact without all the techniques of cell culture: culture of bacteria to clone the gene or to monitor its expression in great number; culture of animal cells to multiply the number of transformed cells;

Artificial Skin by Cell Culture

Skin was the first human organ made in vitro by means of the technique of cell culture. This achievement gives rise to the hope of being able to improve the treatment of extensive burns. Indeed, artificial skin obtained by the culture of healthy skin cells of the patient can be implanted without rejection.

One of the present techniques is to place on the burnt skin an artificial outermost layer of skin, the epidermis. It looks like a transparent veil that protects the organism from dehydration caused by lack of skin protection. In a few weeks, this layer becomes thicker and heals the burnt areas.

Artificial skin is also used for the study of certain diseases and the action of medication used for their treatment. The passage of medication through the skin is also studied to prepare a new way, called transdermic, of giving medication.

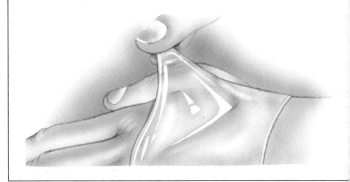

nd plant culture *in vitro* to produce a plant with a new gene.

Other genetic transformations are done on a standard basis; they favor the modification of the genetic material without including the DNA of the cell. It is easy to fuse two animal cells into one. The resulting cell preserves a mixture of the characteristics of the two initial cells. Hence, it is possible to obtain a cell that has both the particular characteristics of cancerous cells, which divide indefinitely, and the characteristics of certain white blood cells that produce specific antibodies. Remember that an antibody is a protein that can recognize an antigen, that is, a molecule foreign to the organism. The antibody binds specifically to the antigen and destroys it. Colonies of such antibodies with unlimited growth are used in the pharmaceutical industry for the manufacture of reactants in biological analysis.

We have learned that plant cells are surrounded by thick rigid walls. These walls can be broken down by specific enzymes extracted from fungi or bacteria. The cell is therefore freed and its membranes provide it with the characteristics of an animal cell. This is called a protoplast to which techniques used for animal cells can also be applied.

For example, now that the cytoplasmic membrane has been bared, two cells can fuse, their genetic characteristics can mix, and new varieties are created. This method is called *somatic hybridization*.

It is furthermore possible to modify the hereditary characteristics of protoplasts. Indeed, certain agents, such as radioactivity, are capable of producing modifications called mutations. A plant regenerated from such a protoplast carries mutations in each one of its cells. When this technique of *somatic mutation* is established, it will enormously simplify the works of plant selecters, that is, producers of new varieties.

In short, cell culture represents the basic technique used in biotechnology. Whatever the transformation may be, it can be amplified by cellular multiplication. This is the great advantage of living organisms.

The Present Forecast of the Future

In the past decade, numerous companies of biotechnology have been founded in the United States, Japan, and Europe. They use cell culture, genetic engineering, and automatic synthesis of macromolecules. The invested funds are high, and many of these companies are controlled by large pharmaceutical, petroleum, and agribusiness industries. The role of international banks appears fundamental.

Present achievements offer a glimpse of the potentials of biotechnology in the fields of agriculture, health, and chemistry: new animals, new plants, new food, new drugs, and new pesticides.

However, a long and exacting job lies ahead because these promises rely as yet on much research to be done in order to become reality.

To improve the genetic quality of their livestock, farmers have been using artificial insemination for a long time. The best bulls are selected, and this method produces at the present time many more offspring than normal reproduction.

Whereas the selection of a superbull was useful, the production of supercows was not until now of any particular interest. Indeed, the number of offspring of a single cow is rather low! The technique of frozen embryos has changed all that.

Toward the Control of Animal Reproduction

The chosen cow is first given hormones that cause multiple ovulation. Thereafter, it is artificially inseminated according

Some agricultural products are already industrial raw materials: wheat, sugar beet, rapeseed, and corn, and their number will increase in the future.

"Custom-made calves." 1. Embryos are preserved in petri dishes at the termperature of liquid nitrogen. 2 and 3. The embryo is divided by minisurgery. 4. A cow can give birth to several calves by transfer of embryos.

to the earlier method. Embryos develop in the cow's uterus, and after a few days they are removed and placed immediately in a freezer at the temperature of liquid nitrogen (-196°C [-353°]), where they are preserved indefinitely.

When needed, an embryo is defrosted and inserted into the uterus of a "surrogate" cow in which the embryonic development is guaranteed. Hence, a "genetic mother cow" that formerly procreated only seven to eight times during its lifetime is now producing hundreds of embryos that are brought to term by surrogate cows.

This number can easily be

oubled if the birth of twins is induced. This is being done by cutting the embryo into two by minisurgery at a very early stage of development.

For the complete control of animal reproduction, it was necessary to be able to choose the sex of the transplanted embryo; male for the butcher, female for the milkman. This became possible very recently.

Thus, we may imagine how the breeder of tomorrow chooses from a catalog the genetic qualities of the father and mother and buys embryos to be carried by cows of lesser performance, and hence cheaper.

Custom-made Plants

The plant kingdom perhaps allows more manipulation than the animal kingdom. Indeed, a plant has a simpler organization. Its transformation therefore appears easier.

Two tools are of major importance: somatic hybridization and genetic engineering. Indeed, we have learned that in plant biology it is possible to reconstruct *in vitro* an entire plant from a single cell. Hence, to act upon a single cell is sufficient!

Somatic hybridization allows the fusion of two protoplasts, that is, cells with their rigid walls removed. After this process, the resulting cell has a mixture of genetic characteristics from two initial cells. This is a very important achievement that in the past was possible only after long experiments of classical genetics that were difficult and expensive. Furthermore, the fusion of cells that do not belong to the same species is also possible. Some years ago, a hybridization between the potato and the tomato, plants of the same family, was achieved in France: the result was baptized the "pomato."

Genetic engineering allows the insertion of a transmitter of a new hereditary characteristic, a gene. Thus several American companies have inserted and monitored the expression of various resistances to herbicides in plants. Experiments in the field are to follow.

To "vaccinate" plants against major bacterial and viral diseases is another future project of genetic engineering.

Numerous and difficult research projects are being carried out to try to produce plants capable of using atmospheric

From the Laboratory to the Factory: Bioreactors

How are microorganisms cultured in great quantity?
Special containers, called bioreactors or fermentors, are used industrially. This is quite different from a culture of bacterial clones in a medium of a few cubic centimeters! The capacity of the largest fermentors, used particularly in breweries, is a million liters. Close to the fermentor, various instruments regulate the temperature and provide air or oxygen, as well as nourish raw materials. Some devices placed in the fermentor, called sensors, "take the pulse" of the fermentation: temperature, acidity, and dissolved oxygen are measured continuously, and data are transmitted to a computer. The latter directs the operation of the fermentor by adapting it to previously established models.

Warning! Fermentation must occur under completely aseptic conditions. Should a parasite microorganism venture by misfortune into the culture medium, it will multiply and contaminate the entire culture. The problem would be the same as happened in breweries at the time of Pasteur.

Industrial fermentors *are so large that they need several phases of culture. Inoculation occurs in fermentor no. 1 (20-500 liters [5-52 gallons]), which is thereafter emptied into fermentor no. 2 (5-20 cubic meters [176-706 cubic feet]), which in turn inoculates the final fermentor no. 3 (50-500 cubic meters [1765-10,540 cubic feet]) whose functions (temperature, agitation, gaseous release, and food) are computer controlled.*

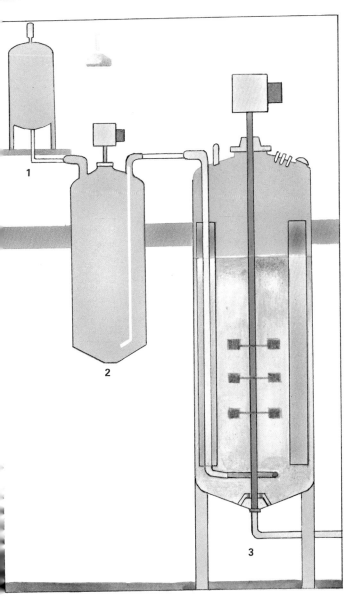

nitrogen to build their proteins. Indeed, today the great majority of plants take nitrogen from the soil in the form of nitrate. Nitrogen is, by the way, the most expensive and most pollutant fertilizer. However, certain plants, such as clover, alfalfa, peas, and alder, do not need these fertilizers. They are exceptions to the rule because they have in their roots bacteria (with a particular metabolism capable of transforming atmospheric nitrogen into organic nitrogen.

Biologists plan to extract the genes of bacteria coding for this performance and to insert them into the DNA of plant cells. Such a plant would not require nitrogen fertilizer, and the savings would be gigantic for worldwide agriculture.

Every 20 seconds, a pump sprays a mist of nutrients in solution on roots that are not in the soil but hang freely from a support of polystyrene.

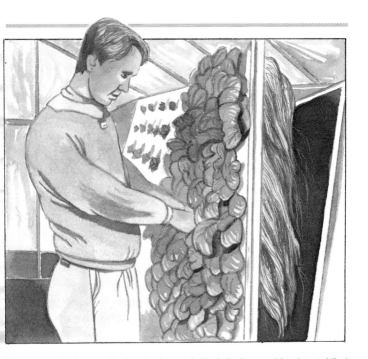

Various types of culture outside the soil are being studied for the development of plants living in difficult environments, such as the desert or in space. This picture shows salad greens attached to a sheet of polystyrene.

A New Gastronomy?

Many scientists are trying to solve the problems of malnutrition in developing countries as well as those of overeating in rich countries. Foods based on soybeans appear of great interest for these purposes, including *miso, tempeh, natto, tofu, sufu,* and *shoyu*! They have been produced in the Orient for the last 2000 years and used as such or as bases for soup, sauces, and spices. Adapted to our food, they might replace meat and milk at a cheap price or be served together with pasta or salads.

But why do we need new foods? Whereas tofu derives from fresh soybeans, all the other products cited derive from fermented soybeans. They are thus of greater nutritional value and contain essential amino acids, vitamins, and mineral salts, but in contrast to products of animal origin, they are low in fat, which in rich countries is the cause of cardiovascular disease.

Against Malnutrition

Today, doctors and dieticians prescribe such food for diets whereas industry is interested in them to produce dietary food for cafeterias, hospitals, and fast food restaurants. Already pâtés and sausages mixed with soybeans are being produced. A new gastronomy is beginning.

In countries where malnutrition is endemic, such food would be of even greater interest. It is of high nutritional value and rich in important proteins. To resort to food of plant origin would save energy, time, and money by avoiding cattle breeding.

Proteins produced from microorganisms could also be used to improve food for animals and humans; these are the PUO, or proteins from unicellular organisms.

When oil was cheap in the 1970s, such proteins created much interest. British Petroleum started to grow cells on diesel fuel. "Petroleum steaks" were going to feed animals. However, when the price of oil skyrocketed, proteins from unicellular organisms became much more expensive than soybeans.

Cell cultures are also considered from other substrates derived from wastes: from dairy industries in France, from sugar beet industries in Cuba, and from forest harvesting in the Soviet Union. In Africa, manioc flour might serve for the cultivation of yeasts that would yield protein enrichment.

The commercial future of PUO remains uncertain. Today, everything depends upon the economy: the price of a barrel of oil, of an oilcake of soybeans, and of traditional food products.

Industries based on flavoring substances, coloring agents, and pectinaceous substances are also part of food biotechnology. It is difficult to predict trends in these fields because for

Food based on plant proteins extracted from soybeans is being consumed more and more. In France, research is being done on other plants such as lupine and wax beans.

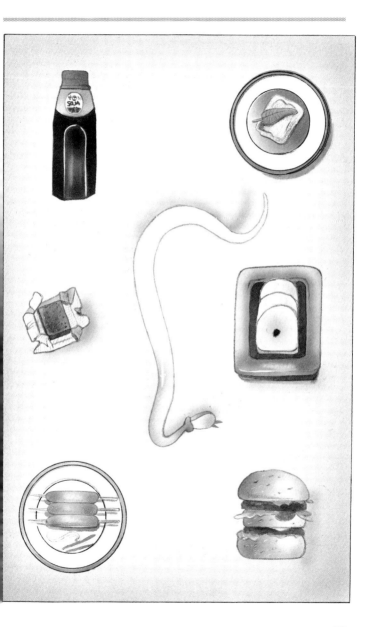

the consumer, ravioli, strawberry yogurt, or ketchup must always have the same taste, odor, and flavor.

Let's take the example of flavoring substances: for some, raw products are used (thyme, fennel, and cinnamon); others are extracts by distillation (pear brandy and cherry brandy); finally, some are synthesized by chemical methods (sweetening, such as saccharine).

The era of biotechnological aromas will start soon. Indeed, microorganisms play an important role during fermentation in giving a taste and an odor to food prepared this way. They produce, for example, some hundred different molecules giving their flavor to beer or to wine.

This is why yeasts and bacteria were cultivated and conversion enzymes were used to manufacture very special aromas: pear-banana, tobacco, or strawberry.

New Molecules for Medicine

Biotechnologies are applied more and more in medicine for hormones, blood, vaccines, and antibiotics.

By means of genetic engineering, a large quantity of a substance present in small doses in an organism can be produced. Such is the case for hormones like insulin, which is injected in the treatment of diabetes to restore the normal ability of the body to use sugars.

Another case exists for a blood factor that controls coagulation (factor VIII). Traditionally, it was extracted and purified from large volumes of donor blood to treat hemophiliac patients, people with a hereditary condition in which the blood fails to clot quickly enough. Today, factor VIII is synthesized in the kidneys of monkeys or hamsters that are manipulated genetically. This technique is less expensive and avoids the risks of contamination.

Genetic engineering is also used in the industrial manufacture of antibiotics.

Since 1940, the date of its first industrial production, penicillin has been made in the same way. First, microorganisms are cultivated in large fermentors; second, the antibiotic is extracted.

Therefore, microbiologists have searched for bacteria and

fungi in the soil to find other useful antibiotics. Thus the following antibiotics were found: cephalosporin, streptomycin, and actinomycin. These molecules can be chemically modified to create new ones.

Genetic engineering offers other new possibilities: increase in productivity of colonies, decrease in production costs, and manufacture of new antibiotics. It remains to be seen if better products than those produced by nature will be made in the laboratory.

Each microorganism has two capacities: one to cause a disease, or *pathogenesis*; the other to trigger the appearance of defense barriers, or *immunology*. Traditional vaccines contain entire microorganisms whose virulence has been greatly lowered. In recently produced vaccines, one finds only part of the microorganism endowed with immunologic capability.

Here genetic engineering can intervene. A vaccine for hepatitis B is produced from genetically transformed yeasts that synthetize the immunogen fraction of the agent of hepatitis B.

Biotechnology, moreover, provides new techniques that are useful in medical diagnosis

Insulin *was the first hormone produced by genetic engineering. It can form crystals, thus indicating its great purity.*

calling for immunology.

Remember that antibodies are proteins that recognize antigens, that is, molecules foreign to the organism. Antibodies bind to antigens and destroy them.

By cell structure, some very pure antibodies are obtained (monoclonal antibodies). They can bind to a "signal molecule" to produce excellent detectors of antigens. The pair thus formed attaches to the site where the antigen is present,

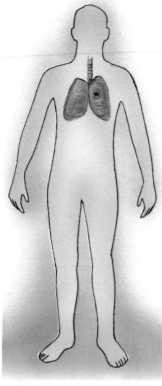

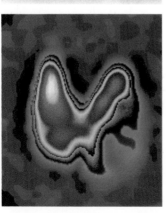

and the signal molecule allows us to find it. This technique is now being used for cancer in order to obtain pictures of tumors. Projects have also begun to attach a toxic molecule to antibodies that destroy the cell of the tumor that carries the antigen.

New Sources of Energy

After the skyrocketing of oil prices, an ancestral source of energy was rediscovered: biomass, the total mass of organic matter consisting of living animals and plants.

To exploit this biomass means to profit from an inexhaustible source of energy: solar energy. Indeed, solar energy is captured by plants during photosynthesis. Therefore, biomass is continuously renewed whereas traditional sources of energy are not, namely, the fossil biomass (oil and coal) and ores (uranium).

*A **cancerous tumor** can be pictured with the help of radioactive monoclonal antibodies, which recognize diseased cells in a specific manner.*

From sugar-producing plants (sugarcane and sugar beet), from cereals (corn, rice, and barley), and from tubers (potatoes, Jerusalem artichokes, and manioc), all rich in starch, the fermentation of yeasts produces ethyl alcohol (or ethanol), which can be used as motor fuel.

Ethanol is used pure or is mixed with ordinary gasoline to form the so-called gasohol already in use in the United States. In Brazil, this motor fuel is made from sugarcane. This fuel provides energy independence, the use of surplus production, and absence of pollution by exhaust gases. In Europe, research on this fuel is being done.

This technique would be of greater interest if wood and all agricultural waste products could be used (such as straw and the branched stem portions of the fruiting cluster of grapes). These materials contain less easily available sugars; however, they are rich in cellulose and in another molecule, lignin, which could be degraded into glucose. To render this physiochemical technique more profitable, fungi could be used because they have great quantities of enzymes that can digest cellulose, whereas left alone in nature, they degrade humus.

"Biogas" also appears to be an interesting source of energy. This gas is released during the fermentation of organic matter by certain bacteria. It is produced in swamps and causes swamp gases.

How to Use Waste Products

Today, household garbage, waste of agribusiness industries, and byproducts of cattle raising are used as substrate thus enhancing the value of 1-2 kilograms (2-4 pounds) of waste produced daily in Europe and in the United States by every individual. During fermentation, this organic matter changes into blackish and granular compost, used as fertilizer.

Biogas provides heating for waste-recycling plants and for certain agricultural buildings (homes, hothouses, and barns for hay storage).

All these methods of waste usage are also excellent means of limiting pollution, but even more methods need to be developed to deal with the increasing problem of pollution.

Oceans and Biotechnology

Oceans are certainly the largest reservoirs of natural resources. Nevertheless, we do not yet know how to exploit them rationally: we are still at the stage of hunting (fishing) and gathering (trawling).

Could biotechnology play any important role?

We must admit that for the time being most attempts are technically not profitable. However, much research is being carried out.

With aquaculture, attempts are being made to breed ocean fish like salmon or perch. Eggs are treated in various ways to accelerate the growth of these fish and to increase their size. These experiments are also underway with crustaceans (shrimps).

Culture of algae is already quite old in the Far East. Research is now being done trying to systematize these methods.

Genetic engineering is used to change certain genetic characteristics that influence the shape and productivity of algae.

Systematic research on many marine organisms (microscopic algae and mollusks) is being carried out to find substances of pharmaceutical interest.

The ocean has also provided models for scientists. For instance, filaments of mussels (byssus) that attach mussels to rocks have led to the invention of a new type of glue that is effective even under water.

However, these attempts remain very limited.

The immensity of the ocean still escapes human conquest.

New Antipollutants

Among the various kinds of pollution, those caused by organic matter are the easist to control by biotechnology.

Everybody remembers the 200,000 tons of oil dumped by the tanker *Amoco Cadiz* on the beaches of northern Finistère (France) more than 10 years ago. It was the first large black tide. The beaches were cleaned by the army and by volunteers.

However, another invisible army took over immediately. Indeed, certain microorganisms are capable of feeding on hydrocarbons, which represent food that is degradable by respiration or fermentation.

Scientists are therefore trying to isolate and modify these microorganisms genetically so that they become good weapons against pollution. Petroleum engineers intend to use them to remove from the effluents of refineries microdroplets of oil that are difficult to recover by mechanical or chemical methods.

However, these microorganisms may play tricks! When they proliferate, they begin to pollute. Thus they may clog up and corrode motors and heaters that function with biodegradable hydrocarbons.

A strange use of microorganisms concerns leaching of min tailings. Indeed, through microbial leaching significant quantities of metals, such as copper, nickel, cobalt, gold, arsenic, and lead, can be recovered. Some colonies of microbes live in rather harsh environments such as volcanic fissures, high temperature hot springs, and iron- or copper-rich rocks. Attempts are being made to use them as microscopic miners.

Are techniques perfected by biotechnology of real commercial interest? We must admit that people have not waited for these techniques to invent mechanical, physical, and chemical techniques for screening, healing, and synthesizing.

Indeed, to reach the market bioconversions, *in vitro* cultures, or fermentations must be more profitable than synthetic products, crops in the field, or manuring, respectively. That battle has to be won by scientists and engineers.

Certain microorganisms feed on petroleum. They are studied for the treatment of black tide (oil slicks).

Promises and Limits

The twenty-first century will certainly see great advances in biology. The achievements of biotechnology will revolutionize biology as much as recent achievements in physics. We are talking here about technical and economic revolutions, but social ones are also starting to be perceived.

Indeed, today's research is posing serious moral or ethical problems because humans have for the first time the capacity to interfere in living beings by manipulating the genetic programs of species. Our environment will therefore be changed accordingly.

Many look into the future with enthusiasm. Let us take, for example, the future of developing countries. Much hope has been put into biotechnology for the improvement of health care and food. Vaccines will be produced in the next years against parasitic, bacterial, and viral diseases that are rampant in these countries. New techniques would allow an increase in the local production of food and the energy output. These countries would thus become more independent. Brazil's manufacture of ethanol, which decreases petroleum imports, is often cited in this respect. With custom-made plants and animals and with home-made energy, most food problems would be solved.

Are these forecasts perhaps hiding other more alarming ones?

An Economic Panorama Turned Upside Down

Commercial ties between developed and developing countries will change. At the present time, the United States and

In a rotating drum, salad greens grow under conditions similar to those that will exist in a space station.

Japan are already partially independent of the sugarcane market because they produce fructose syrup from corn.

The producing countries of sugarcane in Africa, the West Indies, and South America are therefore all the more impoverished.

The same is true for *in vitro* culture of aromatic and pharmaceutical plants, which compete with standing crops of plants imported from countries in warm climates.

Furthermore, biotechnology industries are being merged into a few great multinational consortiums. Moreover, much more research needs to be done before techniques are entirely operational. Their promises are tempting, as we have seen up to now. However, they need large investments of labor and equipment, and only very large companies are able to finance operations with an eye to long-term profits.

Nevertheless, the competition by the consortiums for markets is frantic. For instance, a war of seeds has been declared. The winner will be the company that first puts on the market a plant resistant to a given herbicide as well as the corresponding herbicide.

The farmer will have to buy both at the same time and be commercially dependent upon the producer, the multinational consortium.

To Change the Genome

In the United States, the release in nature of genetically modified microorganisms has met with sharp criticism although the dangers seem to be very small. However, such public reaction has the merit to push for governmental regulations and laws pertaining to all experiments of genetic engineering.

For armament and defense, genetic engineering would be able to produce extraordinarily dangerous bacteria and viruses. The most dreadful war would no longer be nuclear but biological. Fortunately, international agreements were signed or are being signed to prevent this danger.

More serious without any doubt is a transfer of techniques put into practice for animals to humans.

Let us take, for example, the transfer of embryos. After these veterinary miracles, doctors considered the same technique for women.

Indeed, such transfers presented great hope to combat sterility, and the technique is being used successfully in some hospitals.

Normally, the *in vitro* fertilized egg is implanted in the uterus of the genetic mother. However, this is not always possible. In such a case, the fertilized egg is implanted in the uterus of another woman, called the surrogate mother. Today, a woman can carry an embryo that genetically means nothing to her. However, this technique has raised moral, legal, and psychological issues!

The latest possibility concerns the sex of the embryo. Indeed, its determination is part of a total of techniques gathered under the name "prenatal diagnosis," techniques that are very important in medicine. With the help of molecules called probes, a good knowledge of the genetic characteristics of the embryo, its sex, and the screening of serious genetic diseases can be obtained. This early screening allows treatments *in utero*, and for incurable cases, the choice is given to the parents to interrupt pregnancy.

Therefore, prenatal diagnosis sometimes poses difficult psychological problems. To solve these problems, "ethical committees" have been set up in most developed countries. They consist of scientists, doctors, lawyers, top civil servants, and clergy or philosophers concerned with these serious problems.

Genetic engineering applied to humans raises other questions concerning its future application.

In the United States, gigantic mice were recently obtained by genetic engineering. Scientists introduced into the egg cell the gene responsible for the synthesis of a growth hormone. This hormone was produced in such quantity that the animal was born a giant. In the near future, such manipulations will certainly become very frequent. For instance, one can imagine the manufacturing of giant steers for meat.

Genetic engineering has already been used in humans for the therapy of genetic diseases that cause serious retarding effects or deficiencies in the immune system. The gene of the missing enzyme was inserted into those cells that could not synthesize this enzyme. However, genetically modified cells have never been transmitted to offspring because scientists have refused for the time being to touch the reproductive cell.

Similarly, the techniques of fusion and division of eggs have not been used in humans. The first technique consists of a fusion of two fertilized eggs followed by reimplantation in the uterus. Such a technique would result in an embryo with four parents. The second technique divides an egg into two parts, and both would regenerate separately an entire individual and hence produce twins.

Biotechnology in Space

Purification experiments have often provided mediocre results on Earth because of gravity. The latter always introduced irreparable perturbations, microscopic vortices that stir up the medium and remix particles that were separated before.

It is evident that this type of experiment would be of great interest in space, that is, with zero gravity because the cause of perturbations is eliminated.

Electrophoresis, that is, division by means of an electrical field, was practiced during the spaceflight of Apollo 14 in 1971 on its return trip. During the eighth mission of the space shuttle, in September 1983, important steps were achieved: beta cells, producers of insulin, were purified from a mixture of cells from a dog's pancreas. A pure fraction of cells that produce urokinase was also obtained, and various types of hypophysial cells of rats were separated.

In space, the power of division with respect to technology on Earth is increased 700 times and the coefficient of purity 4 times.

Why are these experiments considered important breakthroughs from a strictly biological viewpoint?

Let us not forget that cells separated by electrophoresis can thereafter be produced in great quantities, with the same

Genetic engineering has already produced mice as large as rats. Will they be one day as large as cats?

purity, when placed in culture media. It is possible to consider implants of cells of a specific type. Thus, a deficiency of a hypophysial hormone could be treated by cell implants that secrete this hormone.

Huge productions of particular enzymes can be undertaken commercially, such as urokinase, which is essential in the treatment of thrombosis (blood clots that obstruct a blood vessel).

We understand now why American companies funded the prototype of a pharmaceutical factory launched by the space shuttle. Commercial gains would have been considerable in spite of high costs related to high-level technology because products of great purity and produced in great quantity would eventually have relatively low costs. For instance, a medical treatment of urokinase costing today $1500 would cost only $100 in a few years.

For this reason it is said that biotechnology will develop significantly only in space.

Glossary

Amino acids: Relatively simple nitrogenous molecules that serve as building blocks for proteins. There are some twenty amino acids, and their chains and arrangements provide a specific function for the protein.

Antibodies: Proteins of the immune system that can bind with a particular antigen and having the specific capacity to neutralize the antigen, they help in the defense of the organism (see Antigen).

Antigen: Molecule or fraction of a molecule capable of forming or inducing the formation of antibodies taking part in the immune reaction.

Antisepsis: Fight against pathogenic bacteria.

Aquaculture: Culture of aquatic organisms: algae, shells, fish.

Asepsis: Absence of pathogenic bacteria.

Bacteria: Simple unicellular organisms that have no individualized nucleus.

Biomass: Total mass of living organisms.

Bioreactor: Containment system with optimum conditions for the multiplication of cells or preservation of active molecules (enzymes).

Catalyst: Molecule that either speeds or slows a chemical reaction.

Cell: The smallest biological unit that is independent and can multiply. Some organisms are formed by a single cell: they are called unicellular. The majority of organisms are made up of several, often very numerous cells: they are called multicellular.

Chromosome: One of the bodies of the cell that carries the genes. It is made of a very long molecule of DNA and proteins.

Electrophoresis: In physical chemistry, separation of macromolecules or cells according to their electrical charge.

Enzyme: Protein-based catalyst (See Catalyst).

Enzyme engineering: All the techniques for the best use of enzymatic properties.

Ethics: System of code of morals.

Fermentation: Partial degradation of complex molecules, for instance of food (mostly sugars, such as glucose) in the absence of oxygen. During fermentation, energy used by the cell is released in small quantities (see Respiration).

Gene: Transmittor of hereditary characteristics. It consists of DNA.

Genetic engineering: Programmed insertion of a foreign gene into an organism.

Genome: The total of all chromosomes, the number of which is specific for each gene.

Hormone: Chemical substance formed in glands of the body and transported in the bloodstream to another part of the body, where it has a specific effect. For example, insulin is a hormone secreted by the pancreas.

Hydrolysis: Breakdown of a molecule by the addition of a water molecule. Hydrolysis uses enzymes called hydrolases as catalysts (see Catalyst).

Immunology: Branch of medicine dealing with the defense system of organisms.

In vitro: Literally, what occurs in glass (such as a test tube or a culture dish). Experiments *in vitro* are carried out in an artificial environment, outside the living organism.

Macromolecule: Large molecule.

Microbe (or microorganism): A very tiny living organism, mostly unicellular.

Microbiology: Study of microorganisms and their function.

Mutation: Sudden variation of a hereditary characteristic.

Polymer: A molecule consisting of chains of identical units. For example, starch is a polymer of glucose. Macromolecules are always polymers (see Macromolecule).

PUO: Proteins of unicellular organisms.

Protein: Macromolecule consisting of a chain of amino acids.

Protoplast: Plant cell without its rigid wall.

Respiration: Total breakdown of food (mostly sugars and glucose), in the presence of oxygen. During respiration, energy used by the cell is released in great quantity.

Sex screening: Early detection of an embryo's sex.

Sweetener: Chemical substance with a sweet taste.

Vaccination: Medical technique of stimulating the defense system of an organism by injection of the bacteria causing a certain disease; the pathogenic power of this bacteria, however, has been previously lowered.

Yeast: Microscopic unicellular fungi.

Index

Numbers in italics refer to illustrations.

A

Acids
  amino, 28, 29, *29*
  fatty, 29
  lactic, 14
  nucleic, 23, *27*
Amylase, 32
Antibiotics, *18*, 19, 60, 61
Antibodies
  monoclonal, 61, 62, *62*
Antigens, 61, 62
Antipollutant, 63, 66
Antisepsis, 16, *16*
Apollo 14, 72
Aromas
  biotechnological, 60
Artificial insemination, 50–53
Asepsis, 16
Assyrians, 4
Avitaminosis, 29

B

Bacteria, 9, 10, 15, 16, 18, 19, 24, 25, 26, 44, 46, 47, *47*, 53, 56, 63, 70
Beer, 5, 15–16
Betagalactosidase, 26
Biochemistry, laboratory, *31*
Biogas, 63
Biological war, 70
Biology, 68
Biomass
  fossil, 62
Bioreactor
  enzyme, 35, *35*, 38
Biotechnology
  applied, 10, 26, 40–49, 68, 70–74
  first-generation, 6
  food, 58, 59, 60
  solar, 44, *44–45*, 45
Black tide, 66, *67*
Bread, 4, 6, *6*, 7
Breweries, 4, 5, 15, 34, 54

C

Calmette, 14
Captors, 54
Carbohydrates, 28
Carrel, Alexis, 40
Cell
  egg, 25, *27*, 72
  host, 46
  plant, 22, *22*, 23, 24, *24*, 25, *47*, 49, 53, 56, *56*
Cellulose, 32, 63
Cervoise, 4
Chamberland, 14
Cheese, 7, 10, 12, *12–13*, 13
Chlorophyll, 22
Chloroplast, 22, *22*, 23, 24, *24*
Clone, 25, 26
Cloning, 25, 26, 46, 47, *47*
Coagulation, 60
Coal, disease of, 14
Colony, bacteria, 15
Compost, 63
Corn, 36, *36*, 37, *37*, 70
Culture outside the soil, 57
Cytoplasm, 22, *22*, 23, 24, *24*

D

DNA (deoxyribonucleic acid), 23, 26, *27*, 46, 47, 49, 56
Dextrins, 28
Digestion, 32
Diphtheria, 19
Duclaux, 14, 19

E

Electrophoresis, 72
Embryos
  sex screening, 52, 53, 71
  transfer, 52, *52*, 70, 71
Enzyme
  conversion, 60
  engineering, 30, 32, 33, 34, 35, *35*, 36–37, 38
  "gluttonous," 32
  immobilization, 35, *35*, 37
Ethanol, 68
Ethics, 71
Ethyl alcohol, 7, 14, 63

F

Factor VIII, 60
Fats, see Lipids
Fermentation
  alcoholic, 7, 8, 9, 10, 15, *15*, 16
  lactic, 7, 10, 12, *12*, 13, *13*
Fermentors, industrial, 54, *54*, 55
Fleming, Alexander, 19
Food
  for diets, 14, 58, 59, 60
Fructose, 28, 37, 38

G

Galactose, 25
Gasohol, 68
Gaul, ancient, 5
Gay-Lussac, Louis, 10
Gene, 23, 26, *27*, 46–47, *47*, 49, 53
Genetic engineering, 46–47, *47*, 49, 70–72, *72*
Genetic manipulation, 23, 26, *27*, 46–47, *47*, 49, 52, *52*, 53, 70–72
Genome, 70–72
Glucose, 14, 25, 26, 28, 30, 32, 34, 37, 38
Glycerides, 29
Glycerol, 29
Golgi apparatus, *22*

H

Hebrews, 4
Hepatitis, 61

77

Herbicides, 71
Hooke, Robert, 20
Hormones, 47, 60, 61, 71, 74
Hybridization, somatic, 49, 53
Hydrocarbons, 66
Hydrolase, 32

I

Immunology, 18–19, 61
Insulin, 47, 60, 61, 61
Isomerase, 26, 34

J

Jolly, Justin, 40
Joubert, J.F., 19

L

Lactase, 38
Lactose, 26, 28, 38
Laveran, 14
Lignin, 63
Lipids, 29
Lysosomes, 22

M

Macromolecule, 28, 32
Maltose, 28
Manioc flour, 58
Membrane, cytoplasmic, 22, 22
Meristem, 42, 43
Metchnikov, 14, 19
Mice, 73
Microbes
 pathogen, 16
Microbiology, 10, 11, 14–16, 18–19
Microencapsulation, 35, 35, 39, 39
Microorganism, 10, 11, 12–13, 12–13, 14–16, 18–19, 60
Microscope, 10
Milk, 7, 10
Mineral salts, 30
Mitochondria, 22, 22
Molecule
 signal, 60–62
 toxic, 60–62

Multicellular, 20
Mutation, somatic, 49
Mutations, 49

N

Neuron, 21
Neurotransmittors, 28
Nicolle, 18
Nitrogen
 atmospheric, 56
 organic, 56
Nucleus, 22, 22, 23, 25, 27

O

Organelles
 cytoplasmic, 22, 24

P

Palm, 42, 43
Pasteur Institute, 11, 19
Pasteurization, 15–16
Pasteur, Louis, 4, 6, 10, 11, 11, 14–16, 19
Pathogenesis, 61
Pectinase, 38
Penicillin, 18, 19
Penicillium, 19
Photoreactor, 45
Photosynthesis, 24, 24
Polio, 19
Polymer, 32
"Pomato," 53
Proteases, 34
Proteins
 chemical, 28, 29
 defense, 28, 29
 messengers, 28, 29
 structural, 28, 29
 plant, 59
Protoplast, 53
PUO, proteins of unicellular organisms, 58

R

Rabies, 19
Rachitism, 29
Reproduction, animal, 50–53
Respiration, 15
Reticulum, endoplasmic, 22–23
Ribosomes, 22, 23
RNA (ribonucleic acid), 23, 27
Roehm, Otto, 35
Rosier, 41
Roux, 14, 19

S

Saccharose, 28, 38
Salads, 57, 58
Schleiden, Mathias, 20
Schwann, Theodor, 20
Scurvy, 29
Skin, artificial, 48, 48
Soybeans, 57, 58
Starch, 28, 30, 32, 34, 37, 38
Sterilization, 14–16, 16, 18
 UHT, 15
Substrate, 34, 38, 63
Sugar, 36–37, 36–37
Sugars, see Carbohydrates
Surgeons, 14–16, 16, 18
Surrogate cow, 52, 52, 53
Synthetases, 34

T

Tetanus, 19
Tuberculosis, 19
Tumor, cancerous, 62

U

Unicellular, 20
Urokinase, 74

V

Vaccination, 18–19
Vaccine, 18–19, 68
Vitamins, 29

W

Whooping cough, 19
Wine, 4, 6, 7, 7, 8–9

Y

Yeasts, 6, 7, 16
Yersin, 19

BARRON'S
SOLUTIONS Series

Here's the science series that enables young readers to feel the determination of the world's greatest scientists, as it spotlights the challenges they mastered and the life-changing discoveries they made. Lively chapters and beautiful illustrations provide a clear understanding of how they lived and how their discoveries shaped our future. *(Ages 12 to 13)* Each book: Paperback, $4.95, Can. $6.95, 144 pp., 6" × 9"

ISBN Prefix: 0-8120

MARIE CURIE and the Discovery of Radium, by Ann E. Steinke
One of the most brilliant women ever to enter the world of science, Marie Curie won two nobel prizes. All of her scientific breakthroughs are clearly explained here. (3924-6)

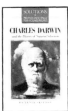

CHARLES DARWIN and the Theory of Natural Selection, by Renée Skelton
Here's the interesting story of how Charles Darwin came up with his Theory of Evolution By Natural Selection, and how it changed the science of biology. (3923-8)

THOMAS EDISON: The Great American Inventor, by Louise Egan
A detailed look at Edison's life and work, this book captures the spirit of one of the greatest inventors of all time. (3922-X)

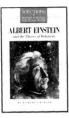

ALBERT EINSTEIN and the Theory of Relativity, by Robert Cwiklik
Albert Einstein changed the science of physics with his amazing Theory of Relativity, and more. This book's clear text explains it all, in simple language that readers will easily understand. (3921-1)

All prices are in U.S. and Canadian dollars and subject to change without notice. Order from your bookstore, or direct from Barron's by adding 10% for postage & handling (minimum charge $1.50, Canada $2.00). N.Y. residents add sales tax.
ISBN prefix: 0-8120

Barron's Educational Series, Inc. • 250 Wireless Blvd. • Hauppauge, N.Y. 11788
Call toll-free: 1-800-645-3476, in N.Y.: 1-800-257-5729 • In Canada: Georgetown Book Warehouse • 34 Armstrong Ave. • Georgetown, Ont. L7G 4R9
Call toll-free: 1-800-668-4336